An account of a Useful Discovery to Distill double the usual Quantity of Sea-Water, by blowing Showers of Air up through the Distilling Liquor

By

Stephen Hales

Double9 BOOKS

An account of a Useful Discovery to Distill double the usual Quantity of Sea-Water, by blowing Showers of Air up through the Distilling Liquor

by Stephen Hales

ISBN: 978-93-59953-38-0

Published by

DOUBLE 9 BOOKS

2/13-B, Ansari Road
Daryaganj, New Delhi – 110002
info@double9books.com
www.double9books.com
Tel. 011-40042856

ABOUT THE AUTHOR

Stephen Hales (17 September 1677 – 4 January 1761) was an English clergyman who made significant contributions to botany, pneumatic chemistry, and physiology. He was the first to take blood pressure readings. He also designed a ventilator, a pneumatic trough, and a surgical forceps for the removal of bladder stones. Aside from these accomplishments, he was a philanthropist and author of a popular book against alcoholic intemperance. Stephen Hales was born in the English town of Bekesbourne in the county of Kent. He was the sixth son of Thomas Hales, heir to the Baronetcy of Beakesbourne and Brymore, and Mary (née Wood), and one of twelve or thirteen children. Sir Robert Hales' grandson, Sir Thomas Hales, 2nd Baronet (Stephen Hales' brother), ascended to the baronetcy in December 1693 since Thomas Hales (died 1692) survived his father, Sir Robert Hales. Hales was schooled in Kensington and Orpington before enrolling at Corpus Christi College, Cambridge (or St Benedict's as it was known at the time) in 1696. Despite the fact that he was a divinity ordinand, hales would have obtained his degree in Classics, mathematics, natural sciences, and philosophy while at Cambridge.

CONTENTS

1. The great Importance of having a sufficient Supply of fresh Water in Ships, has been the Occasion of many laudable Attempts to make Sea-water fresh and wholsome; but all the Attempts and Discoveries hitherto made, have laboured under this great and material Objection, *viz.* the great Quantity of Fuel that was necessary to distill, with a slow Progress, a small Quantity of Water, by any Methods of Distillation hitherto known. But I have lately happily, most unexpectedly, discovered an easy and effectual Method to distill great Quantities of Water with little Fuel; which I was led to by the following Incidents, *viz.* Mr. *Shipley*, Secretary of our Society, *for the Encouragement of Arts, Manufactures and Commerce*, brought me acquainted with Mr. *William Baily* of *Salisbury-Court*, the Author of many ingenious Contrivances; who shewed me, in a small Model of a Tin Vessel, a Method, by which he has happily increased the Force of the Engine to raise Water by Fire, *viz.* by lifting up some of the boiling Water, at every Stroke, by means of a conical Vessel, with small Holes in it, full of Tow; whereby the Quantity of the ascending Steam or Wreak was considerably increased. This led me to think, that a greater Quantity of Liquor might also by this Means be distilled; but on Trial I found the Increase to be only one twelfth Part, tho' considerable in the expanded Form of a Steam. Hence I was led to try what would be the Effect of causing an incessant Shower of Air to ascend thro' the boiling Liquor in a Still; and this, to my Surprise, I found on Trial to be very considerable. There was another Circumstance also, which probably conduced to lead my Mind to this Thought, *viz.* About six Months before, Mr. *Littlewood*, a Shipwright at *Chatham*, came thence purposely to communicate to me an ingenious Contrivance of his, soon to sweeten stinking Water, by blowing a Shower of fresh Air thro' a Tin Pipe full of small Holes, layed at the Bottom of the Water. By this means, he told me, he had sweetened the stinking Bilge Water in the Well of some Ships; and also a But of stinking Water in an Hour, in the same manner as I blew Air up thro' Corn and Gunpowder, as mentioned in the Book on *Ventilators*.

2. The Method, which I used to blow Showers of Air up thro' the distilling Water, was by means of a flat round Tin Box, six Inches Diameter, and an Inch and half deep; which is placed at the Bottom of the Still, on four Knobs or Feet half Inch high, to make room for the Liquor to spread over the whole Bottom of the Still, that the Heat of the Fire may come at it. In larger Stills this Box must be proportionably larger, and have higher Feet. And whereas the Mouth of the Still is too narrow, for the Tin Box to enter, which Box ought to be within two Inches as wide as the Bottom of the Still;

therefore the Box may be divided into two Parts, with a Hinge at one Edge or Side, and a Clasp at the other, to fix it together, when in the Still. This Box must be of Copper for distilling Sea-water; mine was made of Tin for other Liquors also. The Air-pipe, which passes thro' the Head of the Still, will help to keep the Air-box from moving to and fro by the Motion of the Ship; or, if that should not be found sufficient, 3 or 4 small Struts may be fixed to the Sides of the Air-box. They must reach to the Sides of the Still. The Cover and Sides of the Air-box were punched full of very small Holes, one-fourth Inch distant from each other, and about the twentieth part of an Inch in Diameter. On the Middle of the Cover or Lid of this Air-box, was fixed a Nosil more than half Inch wide, which was fitted to receive, to put on, and take off the lower End of a Tin Pipe, which was twenty Inches long, and passed thro' a Hole in the Head of the Still: four Inches of the upper end of this Pipe were bent to a Crook, almost at a right Angle to the upright Stem, in order thereby to unite the Crook to the widened Nose of a Pair of Kitchen double Bellows, by means of a short Leathern Pipe of Calves-skin. See Fig. 1st. This Tin Air-box, and many more of them for other Persons, were made by Mr. *Tedway*, Tinman, over-against the *Meuse-Gate, Charing-Cross.*

3. The double Bellows were bound fast to a Frame, at the upper Part of the Iron Nose, and at the lower Handle, in order the more commodiously to work them. And that the upper Half of the double Bellows may duly rise and fall, in order to cause a constant Stream of Air; (besides the usual contracting spiral Springs within side) several flat Weights of Lead must be layed on the upper Part of the Bellows, near the Handle, with a Hole in their Middle, to fix them on an upright Iron Pin fastened on the Bellows: That by this Means the Weights may the more commodiously be put on or taken off. For, according to the different Depths of the Liquor in the Still, so will the Force of the included Air, against the upper Board of the Bellows, be more or less. Thus, supposing the Depth of the Water in the Still to be twelve Inches; from the Surface of the depressed Water in the Air-Box; then the Pressure of the included Air against the upper Part of the Bellows, will be equal to that of a Body of Water a Foot deep, and as broad as the inner Surface of that Board. It will, therefore, be requisite, to add or take off Weights, according to the different Depths of the Water in the Still, at different Periods of the same Distillation. The Bellows must be proportionable to the Size of the Still, but need not be very large. Wherever the Stills are fixed in Ships, the Air may be conveyed to them from the Bellows, either thro' a small leathern Pipe, distended with Spiral Coiles of Wyre, or thro' Bamboo Canes, or broad small wooden Pipes, like hollow fishing Rods.

4. When I first distilled in this ventilating Way, in order to estimate, what the Difference might be in the Quantity distilled, by that or the

common Method, I tried both Ways, by receiving the distilled Liquor into a Quarter of a Pint Glass, estimating the Times, by a Pendulum beating Seconds. Where I found, to my Surprise, that sometimes three times more was distilled by Ventilation than by the usual Way: But finding Inequalities in the small Quantities thus distilled, in order the more fully and assuredly to ascertain the true Proportion there was in the two Methods of distilling, I put three Gallons of Water into the Still; and, when it boiled, put on the Still-head, and fixed its Nose to the Worm-tub Pipe; which Tub was full of cold Water. When it had distilled for an Hour, the Receiver was instantly taken away. And on measuring the distilled Water, found it to be two Quarts and 45 cubick Inches by a Glass divided into cubick Inches. And a Gallon containing 282 cubick or solid Inches, this Quantity of distilled Water, which was 186 cubick Inches, is 1/1.5 th Part of a Gallon.

5. Then, filling the Still as full of Water as before, and when it began to boil, fixing the Head to the Still and Worm-tub, which was full of cold Water; there was distilled in an Hour, with constantly blowing Showers of fresh Air up thro' the stilling Liquors, five Quarts, less by seven cubick Inches, which is 345.5ths cubick Inches; that is, little less than the double of the Quantity that is distilled in the usual Way. In several other Distillations of a Quart at a Time, I found the Quantity distilled by Ventilation, to be more than the double of that in the usual Way. So that the Quantity by Ventilation, may at a Medium be estimated the double of the usual Distillation. It is the well known Property of moving Air, to carry along with it a considerable Quantity of adjoining Vapour, as also of falling Water to carry much Air down along with it.

6. It is to be hoped therefore, that so considerable an Increase in the Quantity distilled, will be of great Benefit to Navigation, as it may be done in less Time and with less Fire.

7. In the Account of Mr. *Appleby*'s Process, for making Sea-water fresh, which is published by Order of the Lords of the Admiralty in the Gazette of *Jan. 22, 1754*, it is said that a Still which contains 20 Gallons of Water will distill 60 Gallons in ten Hours with little more than one Bushel of Coals; and therefore 120 Gallons in 20 Hours, with little more than two Bushels of Coals. And by Ventilation 240 Gallons, or a Tun; and 24 Gallons may be distilled in twenty Hours, making an Allowance for the times of heating those Stills full of cold Water; and a Still something larger and wider, will distill a Tun in 24 Hours; which will more than suffice for a sixty Gun Ship with 400 Men, whose Provision of Water for four Months is about 110 Tuns. And larger Ships may either have proportionably larger Stills, or else two of them. As for Merchant-Ships with few Men, a small Still will be sufficient.

8. The second sized Stills contain 10 Gallons, and will produce 60 Gallons in 20 Hours, with half the above-mentioned Fuel; and by Ventilation 120 Gallons.

9. The least Stills contain five Gallons, and will produce 32 Gallons in 20 Hours; and by Ventilation 64 Gallons in 20 Hours.

10. I have seen some of these Stills at Messrs. *Steel* and *Stephens*, over-against *Mercers-Chapel*, in *Cheapside*, which have been made for this Purpose. There are Holes in the Feet of the Iron Frame or Stove to skrew them down to the Deck. They were fixed at the Fore-castle before the Mast in King *Charles* the Second's time, when they thought they had discovered the Way to distill Sea-water, free from the noxious Spirit of Salt, and from the nauseous bittern Taste. Or, if it be thought proper, one Part of the Ship's Boiler may be made use of, by adapting a Still-head to it.

11. Now supposing a Still to contain 25 Gallons, and that four Parts in five of it, *viz.* 20 Gallons are distilled off: then, in order to distill *a Tun*, or 210 *Gallons*, the Still must be emptied, cleansed and refilled eleven times; and if the whole be done in 24 Hours, full 16 of those Hours will be taken up in distilling at the rate of a Gallon in about four Minutes and half; and the remaining eight Hours of the 24, being divided into 11 equal Parts, they will be each near 44 Minutes to empty and cleanse the Still, to refill it, and give the Sea-water a proper boiling distilling Heat: whether this can be done in so short a Time, must be known by Experience, and ought therefore first to be tried at Land.

12. Doctor *Butler*, in his lately published Method of *procuring Fresh Water at Sea*, proposes the pouring in more Sea-water into the Still, thro' a Funnel fixed in a small Hole in the Head or Upper-part of the Still, when more than half the former Water is distilled off; by which means the Water in the Still will soon acquire a distilling Heat; and this to be repeated several times; but then it will be requisite to add each time more Chalk, in such Proportion as shall be found requisite. It will be well to try this Method in hopes thereby to increase the Quantity of Water that is distilled. The Hole in the Head, or Upper-part of the Still, is to be stopped with a small Plate of Copper, so fixed as to turn to and fro over the Hole.

13. Doctor *Butler* used capital Soap Lees, in the Proportion of a Wine Quart to 15 Gallons of Sea-water, which sufficed for four or five times repeated Pourings in of more Sea-water into the Still. But as I have found that a small Quantity of Chalk has the same good Effect, and is cheaper, and more easily to be had, it is therefore preferable to Soap Lees.

14. When there is a Fire in the Cook-room, the Sea-water might be ready heated to put into the Still, without any additional Expence of Fuel, in the following Manner, which I shall here describe; tho' I think it probable that it will not be put in practice; yet, as farther Improvements may possibly hereafter be made in it, and as it may be of use in some Cases, at Land at least, I shall here give an Account of it, *viz.*

15. About the Year 1718, Mr. *Schmetou* a *German* Gentleman, got a Patent here for heating great Quantities of Water, with little Expence of Fuel, which he then shewed me. Having fixed a spiral Iron Worm-pipe, in such a Brick Stove or Chimney as Women heat their Irons in, thereby causing the Water to run from a Vessel, thro' the Worm-pipe, several Feet Length round, in the Fire. About 30 Years after I acquainted Mr. *Cramond* of *Twickenham* with this, as hoping it might be of Benefit in distilling Sea-water. Upon which he procured such a spiral Iron Worm-pipe, which was about twenty Feet long, and six-tenths Inch Diameter; the Diameter of the spiral Coile was about fourteen Inches.

16. This I fixed in a Brick Stove in my Garden, with its upper End fixed to a Vessel, which contained 45 Gallons of Water. I found the Event of this first Trial to be as follows, *viz.* When the Water run full Bore, at the rate of a Gallon in 17 Seconds, the Heat of the Water was found, by a mercurial Thermometer held in the Stream, at the lower End of the Pipe, to be 80 Degrees above the freezing Point, 180 Degrees being the Heat of boiling Water. When by means of a Turn-cock, a Gallon of Water was two Minutes in running, then the Heat was 140. At which Rate the 45 Gallons would be an Hour and half in running thro' the Iron-pipe; at which Rate 25 Gallons will run thro' in 50 Minutes, with so considerable a Degree of Heat; and if it was an Hour running, the Heat would approach still nearer to a boiling Heat, when first put into the Still, which would forward the Distillation if wanted.

17. I pumped the heated Water up again into the upper Vessel; and thus continued to circulate the heating Water, till its Heat was 160 Degrees in the upper Vessel, *viz.* within 20 Degrees, or one-ninth of boiling, the Heat requisite for plentiful Distillation. I was in hopes that if the Water in the upper Vessel could have been brought to a due Degree of Heat, and a Still-head were fixed on it, with its cooling Worm-tub, then Water might have been distilled in Ships, by having the Iron Worm-pipe fixed in the Chimney of the Cook-room: But I found, that when the Heat of the Water in the upper Vessel was 160 Degrees, *viz.* within one-ninth of boiling; then in running through the Iron Worm-pipe again, it was so over-heated, as to expand in the Pipe, into an explosive Vapour, which hindered the running

of the Water. However I thought it not improper to give an Account of this Attempt, notwithstanding it failed. Not knowing whether this Method of heating Water, may not in some Cases, at Land at least, be of use, thereby to save, in some degree, both Fuel and Time: Perhaps an Iron Worm-pipe of a larger Bore might do better.

18. The Waste of Fuel will be less in proportion to the Quantity distilled in large, than in small Stills; and the wider the Still-head is, so much the more Liquor will be distilled, and more with a Worm-tub than without it. The Worm-tub may be so covered, as to prevent the flowing over of the Water by the Motion of the Ship.

19. It is of great Importance to take care to keep all Parts of the Still clean, that there may be no Rust or Verdigrease in the Copper, which will occasion Vomiting.

20. If it be necessary, the better to close the Joining of the Still-head, it may be done with a Lute or Paste made of a Mixture of powdered Chalk and Meal, wetted with Salt-water.

21. Now that several effectual Means are discovered, to make distilled Sea-water wholsome, and also to distill it in much greater Quantity in the same Still, in the same Time, and with nearly the same Quantity of Fuel; it is reasonable to believe, that it will be of great Benefit to Navigation, not only in saving much stowage room, for other important Purposes; but also in procuring fresh sweet wholsome Water, instead of stinking putrid Water, hitherto used; which must needs have a Tendency to promote that putrid Distemper, the Scurvy. And if withal due Care be taken, to exchange for fresh Air the putrid close confined Air of Ships, which has occasioned the Death of Millions of Mankind; then Navigation will become remarkably more healthy, and with little more Danger to Health and Life, than at Land, except from Storms.

22. Now supposing, that in a sixty Gun Ship, the 110 Tuns of Water, for four Months use, were distilled at the Expence of three Bushels of Coals to a Tun, this would consume nine Chaldrons of Coals: And as a Chaldron of Coals weighs about a Tun and half; hence it appears that Coals will distill about eight times their Quantity of Water. And the 110 Tuns of Water weighing (at the Rate of 2240 Pounds to the Tun) 138 Tuns; and the nine Chaldrons of Coals weighing thirteen Tuns and half, that is 94 Tuns and half less than the 110 Tuns of Store-water; and allowing twenty-four Tuns and half for the Still, Water-casks, and Coals, there will be 70 Tuns Weight of Stowage saved thereby for *other Uses*. Or if some Tuns of Store-water are carried by way of Precaution, which it will be advisable to do, especially at first, till they can be assured, by repeated Experience, what Quantity can be

depended upon by Distillation; even then about half the Tunnage will be saved, which will be a very material Advantage.

23. Tho' when the distilling Liquor runs from the Bottom of the Worm-pipe, thro' a long Pipe fixed to it, the Waste by the ventilating rushing Air, is not great when the Water in the Worm-tub is not hot; yet the following Precaution, if needful, may be used, in distilling by Ventilation, *viz.* to fix at the lower End of the Worm-pipe, by means of a wooden Fawcet, a small Cask for a Receiver; the Fawcet to enter the upper side of the head of the Cask, and in order to give a free Passage for the great Quantity of ventilaing Air to pass off, and withal at the same time to prevent the escaping of much moist Vapour with it, it will be proper to fix at the Bung-hole a long upright Pipe of Wood, or of any Metal. I used a Gun-barrel four Feet and a half long; through which some small Degree of moist Vapour escaped; as appeared by the Dampness of a Piece of Paper, fixed at a little Distance above the Mouth of the Gun barrel. This Vapour became visible, and much increased, when the Water in the Worm-tub was very hot; at which Time, less is distilled into the Cask-receiver; then also there is more Danger of the Spirit of Salt arising. And it was observable, that the Water in the Worm-pipe Vessel heated much sooner by Ventilation, than in the common Way of distilling. For which Reason that Water ought to be changed so much the oftener, which can easily be done at Sea. The Cocks also at the Side of the Worm-tub ought to be large, in order to let the hot Water off the faster.

24. But tho' the Water in the Worm-tub was sooner heated by Ventilation, because a double Quantity of hot Steam passed thro' it, more than passed thro' it in equal Times in the common Way of distilling; yet in the usual Way of Distillation the Liquor in the Still is hotter, with equal Fire, as is evident by its aptness to boil over thro' the Worm-pipe; whereas in the ventilating Way it did not boil over, notwithstanding a very hot Fire was purposely made for a Trial. The continual Streams of ascending fresh Air, not only in some Degree abating the Heat of the Water; but also incessantly carrying off the more rarefied Particles of the Water, which, when expanded into a repelling State, do thereby cause the overflowing Ebullition of the Water. On which Account it is probable, that less Spirit of Salt is formed and raised by Ventilation than without it. As also on account of the fresh Air ascending, not from the Bottom of the Still, where is the greatest Plenty of Salt, especially towards the latter End of each Distillation; but about three Inches from the Bottom, *viz.* principally from the many Holes at the Surface of the Air-box.

25. And whereas the Quantity raised from the Still, and distilled into the Cask-receiver, cannot be seen; the proper Quantity to be distilled in each Distillation, may with great Accuracy be known, by having a well closed

Pewter Bottle of the Size of about half a Pint, with a Brass Wyre as big as a Goose Quill fixed to it, the Wyre to pass thro' the Receiver-cask, near the Bung-hole, which the floating Pewter Bottle will raise up, till the Marks on the Wyre appear just above the Cask. I made use of a Glass Viol for this Purpose. This Wyre will rise and fall freely, notwithstanding the Motion of the Ship, if it passes not only thro' the Wood of the Cask, but also thro' a metaline Pipe two or three Inches long, fixed in that Hole. And it will be known by the simmering or boiling Noise of the Water in the Still, whether it is hot enough to distill; for the running of the Water into the Receiver-cask cannot be seen.

26. As it might be suspected, that more Spirit of Salt would be raised, and distilled over in the ventilating Way, than without it; having procured 18 Gallons of Sea-water by the *Margate* Hoy, which was taken up at some Distance from the Shore, I put three Gallons of this Sea-water, as soon as I had received it, into the Still; and when it began to distill, Air was blown up thro' it. For some Time, as is usual, in the Distillation of Sea-water; no Spirit of Salt arose; but after distilling some Time longer, there were very weak whitish Clouds, with Drops of Solution of Silver in Aqua-fortis, as in the common Way of distilling. Hence we see, that Ventilation does not increase the Quantity of Salt, but rather probably somewhat decreases it, for the Reasons above given, No 24.

27. I distilled three Gallons of Sea-water, which had stunk and became sweet again; when about ten Quarts of it had been distilled off, then there began to be very weak whitish Clouds with Solution of Silver, but none with Solution of Mercury; which shows the Water to be hitherto good, agreeably to what I formerly had found to be the good Effect of distilling Sea-water, which had putrified, and become sweet again; of which I published an Account in the Year 1739. But when I continued the Distillation on, a quarter of an Hour longer, *viz.* till there was but a Pint of Water remaining in the Still, and the Salts were incrusted on its Sides, up near three Inches from the Bottom, and lay in Heaps at the Bottom of the Still, then the distilled Liquor had whitish Clouds in it, with the Solution of Mercury in Aqua-fortis. From this Distillation we see, that Putrefaction, by dissolving the bittern Salt and Bitumen, into very minute Parts, qualified them to combine with the more fixed common Salt, so as to detain them from rising in Distillation.

28. I distilled three Gallons of Sea-water, with the Proportion of six Ounces of Mr. *Appleby*'s Lapis Infernalis, and six Ounces of calcined Bones to 20 Gallons of Sea-water, as he directs. This Water lathered well with Soap, and boiled Peas well.

29. I distilled also some Sea-water With half an Ounce of Stone Lime to a Gallon, from the *Clee* Hills in *Herefordshire*, which having been preserved ten Months in a Firkin, had slacked to dry Powder. This distilled Water did also lather well with Soap, and boiled Peas well; which proves that the Lime, which is a fixed Body, does not distill over with the Water. Since I made this Distillation, General *Oglethorpe* informed me, that his Father, Sir *Theophilus*, told him, that Lime was one of the Ingredients, of what he and the rest of the Patentees, in *Charles* the Second's time, called the Cement, with which they made distilled Sea-water wholsome.

30. I distilled also some Sea-water with the like Proportion of powdered Chalk, which boiled Peas well, and was better tasted than the Waters distilled with Lapis Infernalis or Lime. I distilled also some Sea-water with an Ounce of Chalk to a Gallon, but found no Difference in the Taste of this, and that which had but half an Ounce of Chalk to a Gallon: So that half an Ounce of Chalk to a Gallon of Water will be sufficient; but where the Sea-water is salter, or more bituminous, more Chalk may be added if needful.

31. Dr. *Alston* of *Edinburgh*, in the Preface to the Second Edition of his Dissertation on Quick-lime and Lime-water, says, that "the like Effect was found in distilling Sea-water with Lime, that it neither precipitated a Solution of Silver in Aqua-fortis, nor a Solution of corrosive Sublimate in Water, nor did it form a Pellicle of various Colours on its Surface, as did the Water distilled by Mr. *Appleby*'s Process." And I find, Page 35 of my Book on this Subject, that Lime of Oyster-shells had the same good Effect, but required two Distillations; I might then use too small a Quantity of that Lime. Hence it is probable, that the Chalk, the Lime, the Lime in the Lapis Infernalis, and the Lime in Dr. *Butler*'s Soap-lees, seize on and fix not only the bittern Salt, but also the Bitumen of the Sea-water, as we learn from the like Effect in the Purification of the Salt of Hartshorn. That the saline Spirit arises chiefly from the bittern Salt, and not from the more perfect Sea-salt, is probable from hence, *viz.* when I distilled three Gallons of common Water, made as salt as Sea-water with common Salt; no Spirit of Salt arose, even tho' the Distillation was carried so far as to leave the Salt, tho' very damp, to lie in Heaps, and it was incrusted on the Sides of the Still, for about three Inches from the Bottom.

32. It is a considerable further Advantage, that Water thus distilled by Ventilation, being thereby repleat and freshened with Air, has for present Use a more agreeable Taste, than Water distilled without Ventilation, which requires the standing a longer Time to have its more disagreeable adust Taste go off. And as the volatile Oil of Pepper-mint does rise on the Wings of the ventilating Air during the Distillation; so also may that Part of the

Bitumen, which is volatilized by Heat; as also the volatile urinous Salts of the Sea-water, which arises from animal Substances, be sublimed in the same Manner.

33. It was observable, that the Water distilled fast, even tho' the Water in the Still was below the Surface of the Tin Air-box, thro' which the greatest Part of the ascending Shower of Air rushed. Hence the ventilating Air, in ascending among the Vapours, carries them off fast. Hence it is to be suspected, that this Method of Ventilation will not do well for simple Waters, or fermented vinous Spirits; because they being very volatile, much of them may be carried off in Waste.

34. It was observable, that in these Distillations of Sea-water, no whitish Clouds appeared on dropping in Solution of corrosive Mercury, not even when considerably more than four Parts in five of the Water had been distilled over. And it was the same with the Mixture of Lapis Infernalis, Lime, and Chalk; whence it is probable, that the Lime and Chalk seize on and fix the more volatile bittern Salt, as does also the Lime in the Lapis Infernalis. And it is well known, that Sugar, that sweet Salt, cannot be made without Lime, on which, as its Centre of Union, it fixes and granulates.

35. And whereas with a Solution of Silver in Aqua-fortis, which was much weakened and diluted with Water, there appeared a faint Degree of whitish Cloud, in all the above-mentioned Distillations, tho' not with the stronger Solution of Mercury till the Distillation was carried on, much beyond four Parts in five of the Water in the Still; when both Solutions caused remarkably white Clouds, especially the Solution of Mercury; which indicates the Quantity of the Spirit of Salt which was raised during the former part of the Distillation to be exceeding small, since it could not seize on, nor disengage the Aqua-fortis from the stronger Solution of Mercury, tho' it did in a very small Degree in the weak Solution of Silver, so as to let loose a very little of the Silver, which thereby caused the faint Clouds. When a Drop of the Solution of Mercury was dropped into the distilled Water, after a Drop of the Solution of Silver, it resorbed the Silver Cloud, and made the Water clear, by means of the great Proportion of acid Aqua-fortis that was in it.

36. Now in order to make some Estimate of the very small Quantity of Spirit of Salt in these several distilled Waters, I dropped a Drop of the Solution of Silver into an Ounce, or 480 Grains of pure Rain Water, which gave no Clouds; but on dropping in a Drop of Sea-water, which weighed a Grain, the white Clouds were strong. And since Sea-water can dissolve nine times more Salt than it has in it; therefore, supposing the Drop to be so fully impregnated with Salt, then the Salt would be the 480th Part of the

Ounce of Water. But as there is nine times less Salt, therefore the Proportion of the Quantity of Spirit of Salt will be but the 4320th Part. And how much less must be the Proportion of Salt in these distilled Waters, which is not sufficient to make a sensible Impression on Solution of Mercury, and but a faint one on much diluted Solution of Silver. Such distilled Sea-water will not therefore, probably be unwholsome; almost all Spring-waters have some Degree of Salt in them: But if there were more of the Spirit of Salt, a very small Quantity of Pot-ash, or Pearl-ashes, or Salt of Tartar, combined with it, will turn it into common Salt, the Quantity of which would be extremely little.

37. It may be well to be provided in Ships with some Silver dissolved in Aqua-fortis, mixed with pure Rain-water, or distilled fresh Water, in the Proportion of sixty Drops to an Ounce of the Water; tho' it is probable, it may seldom be wanted, unless in some doubtful Cases, when the Taste may not be accurate enough to perceive, whether there be any Spirit of Salt in the distilled Water.

38. Since double the usual Quantity of Vapour may by Way of Ventilation be carried off, common Salt may thus be made much sooner, cheaper, and better; because as there is much less Fire used; so proportionably, less of the fine acid Spirit of the Salt, in which its Virtue consists, will be evaporated away: For it is well known, that the Salt is best which has undergone the least Action of Fire in making.

39. This more speedy Method of evaporating will also be useful, in making many other Evaporations; as in making Pot-ash, *&c.*

40. But some are apprehensive, that this great Improvement in Distilling, may be of ill Consequence in making those destructive Spirits cheaper, which are already but too cheap. Had not the Improvement been of great Benefit to Mankind in many other Respects, I should have been far, very far, from endeavouring after it, or discovering it. But should the Event be to make those Spirits cheaper, and consequently, by spreading farther, more destructive, the consequence of that will be, that the increased raging Devastation will the sooner necessarily rouse the Nations to put a Stop to what must be done hereafter; for if the Ravages continue increasing, as they have done for sixty Years past, the human Species must needs not only be greatly debased, but even in great measure diminished and destroyed. And yet none of the Nations, whose very Vitals are thereby consuming and destroying, endeavour to put any Stop to it, except the Heads of the native Indians in *North-America,* who have long repeatedly intreated the *English* to sell them no Rum; which is as effectually extirpating of them, as the Hornet did the unsubdued remainder of the *Canaanites*.

41. If Mankind, instead of receiving and entertaining this Pest with almost universal Applause and Approbation, could prevail with themselves to be in earnest to use Means to deliver themselves from it; then much might be done towards it, by lowering and weakening all kind of fermented distilled Spirits with Water, to a salutary Degree, as is now practised in our Plantations in *America*, in making Punch so weak, as not to be hurtful; which, when it was much stronger, was well known to destroy Multitudes. And where the like humane, wise, and laudable Practice has been used in Ships, it has had the same happy salutary Effect.

42. What Necessity or even Temptation can there be to be averse to the making them wholsome, instead of being venomous and destructive? and that not only of the Lives, but even of the Morals of Mankind. How much therefore does it behove all, who have any Concern for the Honour and Dignity of their own kindred Species, any Indignation at its being thus debased and disgraced, any Bowels of Pity for the vast Multitudes, not less than a Million, that are yearly destroyed all over the World, by this moral as well as natural, and therefore worst of all Evils that ever befel unhappy Man; to use their utmost Endeavours to deliver Mankind from this Pest? But notwithstanding this astonishing Ravage and Destruction of the human Species, yet the unhappy unrelenting Nations of the World, seem as unconcerned about it, as if only so many Thousands, nay, Millions of Caterpillers or Locusts were destroyed thereby. Was there ever a more important Occasion to rouse the Indignation of Mankind? Can we be calm and undisturbed, when this mighty Destroyer rears up its invenomed Head every where? The most zealous Advocates for Drams, even the unhappy besotted Dramists themselves, the prolonging of whose Lives, and whose real Welfare both here and hereafter, is hereby sincerely intended, cannot find fault with this well-meant Remonstrance, in Defence of them, and of all Mankind, against this mighty Destroyer, from one who has long been labouring, and that not without Success, in finding Means to preserve Multitudes of Lives, by various Ways.

An Account of the great Benefit of Ventilators *in many Instances, in preserving the* Health *and* Lives *of People, in Slave and other Transport Ships.*

43. It is to be hoped that the several Means here proposed for having fresh and sweet Water at Sea, will be of great Benefit in preserving the Health and Lives of Multitudes of that valuable and useful Part of Mankind, those who occupy their Business in great Waters; whose Welfare I have long had at heart, and endeavoured to promote by various Ways; especially by finding Means to procure them fresh salutary Air, instead of the noxious, putrid, close confined pestilential Air, which has destroyed Millions of Mankind in

Ships. And it is to be hoped that by diligent Researches, farther and farther useful Discoveries will hereafter be made for the Benefit of Navigation.

44. The following, as they are strong Proofs of the great Benefit and Usefulness of Ventilators in Ships, so they also fully prove that they can most commodiously be fixed and worked in them, in contradiction to the vulgar, false, and groundless Notion, that they take up too much room, and are incommodious, and in a manner impracticable to be worked, whereas the Men are eager to work them; and many more Persons can be with Safety to their Health and Lives in a ventilated, than in an unventilated Ship; which fully obviates the Objection as to the Room they take up. In new and important Researches, the likeliest Way to succeed, is to pursue a Thought not only by imperfect and fallacious Reasonings, but when the Nature of the Thing requires it, with a proper Series of Trials and Experiments. Thus in the present Case, the principal Cause of the Sickness in Ships, is the noxious putrid Air; the obvious Remedy is the exchanging that foul Air for fresh, by effectual Means, which are seldom discovered by dwelling only on Objections, but are usually the Reward of repeated diligent, experimental Researches. Neither are we to be discouraged in these our Pursuits by some Disappointments, for I have frequently found that they lead to the Thing sought for: And by the like Clue of Reasoning and Experimenting, there is the greatest Probability that we shall succeed in another very important Research, *viz.* the preserving much longer from Decay the Timbers of Ships laid up in ordinary in Harbour: For as we are assured by daily Experience, that the Decay is wholly owing to damp, close confined putrid corroding Air; so the only Remedy for this Evil, is the frequently changing the Air among the Timbers, by plentiful Ventilations; which we find by happy Experience, can be effected to such a Degree, as give reasonable Hopes, enough to encourage our farther Trials and Researches.

45. Captain *Thomson* of the *Success Frigate,* in his Letter to me dated *London, Sept. 25, 1749*, says, "That during the Ventilation, the Lower-deck Hatches were commonly kept close shut; by which means the Air was drawn down into the Hold, from between Decks, thro' the Seams of the Ceiling, along the Timbers of the Ship; by which means we found the foul Air soon drawn off from between Decks. Our Rule for ventilating was for half an Hour every four Hours; but when the Ventilating was sometimes neglected for eight Hours together, then we could perceive, especially in hot Weather, a very sensible Difference by that short Neglect of it; for it would then take a longer Time to draw off the foul Air. Our general Rule was, to work the Ventilators till we found the Air from them sweet. We all agreed that they were of great Service; the Men being so sensible of the Benefit of them, that they required *no driving* to work that which they received

so much Benefit by. We found this good Effect from Ventilation, that tho' there were near 200 Men on board, for almost a Year, yet I landed them all well in *Georgia*, notwithstanding they were pressed Men, and delivered me out of Goals, with Distempers upon them. This is what I believe but few Transports, or any other Ships, can brag of; nor did I ever meet the like Good-luck before; which, next to Providence, I impute to the Benefit received by the Ventilators. It is to be remarked, that we who lay wind-bound, for four Months, with our Expedition Fleet, which soon after invaded *France*, were very healthy all the time, when they were very sickly in all the Ships of that Expedition.

46. "This certainly occasioned all kind of Grain Provisions to keep better and longer from Weevels, than otherwise they would have done; and other Kinds of Provisions received Benefit from the Coolness and Freshness in the Air of the Ship, which was caused by Ventilation."

47. Mr. *Cramond* also informs me, that he found the good Effect of Ventilators on board a Slave-Ship of his with 392 Slaves, twelve of which were taken on board, just before they sailed from *Guinea*, ill of a Flux, which twelve all died; but the rest, with all the *Europeans* in the Ship, arrived well at *Buenos Ayres*.

The following is a Letter to me from Captain ELLIS, *viz.*

"Sir,

48. "Could any thing increase the Pleasure I have in a literary Intercourse with you, it would be to find that it answered your End in promoting the public Good. The *Vis-inertiæ* of Mankind is not the only Difficulty you have had to encounter, but their Ignorance and Prejudices, which are almost insuperable. It is to your Perseverance and Resolution, that the little Progress you have made is due: Indeed I ought not to say little; for it is a great Step to have found the few that have Hearts good enough to relish your Plan, and Heads sufficiently clear to discern the most effectual Method of advancing it. It does Honour to those noble and other worthy Personages that join you in Acts of such extensive Humanity, as the Introduction of Ventilators to Hospitals, Prisons, Ships of War and Transport, *&c.* as they must necessarily render the Miseries of the first more supportable, and the close and constant Confinement of the others less prejudicial and fatal to their Health and Life. It is to be lamented that they are not more generally made use of; for, notwithstanding their Advantage is apparent and incontestable, it is scarce credible how few are to be found among the vast Number of Ships daily employed in carrying Passengers, Slaves, Cattle, and other perishable Commodities. Those of your Invention, which I had, were of singular Service to us; they kept the Inside of the Ship cool, sweet,

dry, and healthy: The Number of Slaves I buried was only six, and not one white Man of our Crew, (which was thirty-four) during a Voyage of 15 Months; an Instance very uncommon. The 340 Negroes were very sensible of the Benefits of a constant Ventilation, and were always displeased when it was omitted: Even the Exercise had Advantages not to be despised among People so much confined. I must not, however, forget that Ventilation alone is insufficient to keep Disorders out of Ships; for often Infections are brought aboard by the Slaves, or others; and frequently Diseases are produced by feeding on bad or decayed Food, but oftener still by Insobriety; for I have ever remarked that the immoderate Use of spirituous Liquors in warm Climates, is more pernicious and fatal even than the Malignancy of the Air itself. In cold Countries too, where I have had Experience, those Sailors, or others, who accustomed themselves to hard drinking, especially of Drams, had the Scurvy in a terrible Degree; whereas those who were temperate or sober, either escaped it entirely, or had it but moderately. The Effects of Drunkenness was still more discernable among the Indians adjoining our Settlements in *Hudson's-Bay*, who are a feeble, diminutive, chilly, indolent Set of People. On the contrary, those who come from the inland Parts (who are unused to drink Brandy) are brave, active, robust, and industrious. The same Difference is observable in the *Africans*, and perhaps among the Inhabitants of most other Nations, did we attend to it. It was to the unusual Sobriety of my Crew, that I ascribed, in some measure, their uncommon Healthiness; for Sailors breathe a purer Air, and enjoy more Exercise and Liberty, than Passengers or Slaves; wherefore their Ailments are owing to bad or disorderly Living, as well as to unwholsome Air.

"Could I but see the immoderate Use of spirituous Liquors less general, and the Benefits of Ventilators more known and experienced, I might then hope to see Mankind better and happier. I am,

"SIR,
"Your most obediens Servant,
HENRY ELLIS."

Bristol,Dec. 26, 1753.

49. And, by the like good Conduct, in his next Voyage in the Year 1755, not one of 312 Slaves died; and all his 36 Sailors arrived alive and well at *Bristol*.

50. And the Earl of *Halifax* has often informed me of the great Benefit they found by the Use of Ventilators, in several *Nova Scotia* Transport-Ships, twelve to one more have been found to die in unventilated than in ventilated Ships. It is indeed a self-evident Thing, that the changing the

foul Air frequently in Ships, in which there are many Persons, will be a means of keeping them in better Health than not doing it; which makes it the more astonishing that effectual Proposals to remedy so great an Evil, should be received with so much Coldness and Indifference by Mankind. They little consider that it is the high Degree of Putrefaction (that most subtile Dissolvent in Nature) which a foul Air acquires in long stagnating, which gives it that pestilential Quality, which causes what is called the Goal-Distemper. And a very small Quantity, or even Vapour of this highly attenuated Venom, like the Infection or Inoculation for the Small-pox, soon spreads its deadly Infection. Ought not Men therefore, from the common natural Principle of Self-Preservation, to use their utmost Endeavours to shun this pestilent Destroyer, by which Millions of Mankind have perished in Ships?

An Account of some Tryals to cure the ill Taste of Milk, which is occasioned by the Food of Cows, either from Turnips, Cabbages, or autumnal Leaves, &c. Also to sweeten Stinking-water, &c.

51. This Method of blowing Showers of Air up thro' Liquors, will be of considerable Use in several other Respects, as well as in Distillation, as appears by the following Trials, *viz.*

52. I have been informed that it is a common Practice, to cure the ill Taste of Cream from the Food of Cows, by setting it in broad Pans over hot Embers or Charcoal, and continually stirring it, till scalding hot, and till cool again: But when I attempted to do this much sooner, and more effectually, by blowing Showers of Air up thro' it; I soon found it to be impracticable, by reason of its very great Degree of frothing up. The ill Taste must therefore be got out of the Milk, before it is set for Cream; which I have been told, has been practiced, and that with some benefit, by giving the Milk a scalding Heat, without stirring it.

53. *May* 22. I ventilated some ill tasted, new unheated Milk of a Cow which was purposely fed with Crow-Garlick mixed with cut Grass. After 15 Minutes Ventilation the Taste was a little mended; in half an Hour's blowing it was something better. At the Hour's end it had the same Taste, but was sensibly better than the unventilated Milk. I was disappointed of an Opportunity to repeat the Experiment with Crow Garlick Milk, with a scalding Heat; it would then probably have been soon perfectly cured; as it is reasonable to believe from the Event of the following Experiments, *viz.*

54. *August* 23, four Quarts of ill tasted new Milk, from a Cow which had fed eighty-four Hours on Cabbage Leaves only, and drank during that Time very little Water; were put into a leaden Vessel, eight Inches in Diameter,

and thirty Inches ches deep. The leaden Vessel was heated in a large Boiler, and set into a Vessel of hot Water; thereby to give the Milk a scalding Heat, and also keep it hot. In ten Minutes Ventilation it was perfectly cured of its ill Taste; and after standing twenty-four Hours in a broad Pan, there was a thick Scum which was half Cream and half Butter, free from any ill Taste; the skimmed Milk was not sheer or thin: So here is a Method to make good Butter from ill tasted Milk.

55. The Froth of the Milk was so great, by reason of a too brisk Ventilation, as to make it froth over the Vessel, which was thirty Inches deep; if it had not been kept down, by constantly lading and breaking the very large Bubbles of Froth. But when the Ventilation is more gentle, the Froth has risen but three Inches from six Quarts of Milk, which was nine Inches deep. The Cabbage Milk was but six Inches deep. I repeated the like Operation the same Day, with the Evening Milk of the same Cow; but giving it only a Heat, that I could bear my Fingers in, for a little Time; with this Degree of Heat, after forty-five Minutes Ventilation, the Milk (tho' much better tasted) yet was not so compleatly cured, as the former Milk. Hence we see, how necessary Heat is, to volatilize the rancid Oyl (which gives the ill Taste) to such a Degree as to cause it to fly off by Ventilation.

56. It was observed that what was milked from this Cow a Week after she had done eating the Cabbage, had an ill Taste.

57. I have not as yet had an Opportunity, to try to cure, in the same Manner, the ill Taste of Milk, which is occasioned by Cows feeding on autumnal Leaves, or Turnips, they having probably eaten this Autumn, the fewer Leaves, on account of the Plenty of Grass, occasioned by much Rain; which has also hitherto prevented Turnips from being rancid, which are observed to be most so, when they shoot out in the Spring. As Opportunities offer I purpose to make Trials, which I conclude others will also do, which will probably be attended with the same good Effects as that on the Cabbage Milk.

58. But tho' the ill Taste of Milk from feeding on Cabbage Leaves, was thus effectually cured by volatilizing with Heat, and dissipating by Ventilation the rancid Oil; yet the bitter Taste of a strong Infusion of Chamomel Flowers in six Quarts of Water, was not sensibly abated by an Hour's Ventilation of it, while scalding hot.

59. I am informed that, in *Devonshire*, they set the Pans of Milk on Trivets, making Fires under them, to give the Milk, gently and gradually a scalding, but not a boiling Heat, which would disturb the rising Cream; and then set it on the Floor in the Milk-house to cool, where in twelve Hours it has a thick Scum, partly Butter and partly Cream: The skimmed Milk is very

thin and sheer; and the Cream in great Plenty and delicious, except it gets a smoky Taste, which it is apt to do; and which might probably be prevented, by having a Range of as many Stoves, as there are Pans of Milk to be used at one Time; all to be warmed by one Fire, either at one end, or the middle of the Flue or Funnel in the Brick-work, which conveys the Smoke and Heat under the Stoves. And as the Pans nearest to the Fire will soonest have their due Heat, on their Removal to bring the farthest and coolest Pans nearest the Fire; and instantly covering the uncovered Stoves with proper Covers to prevent the Heat and Smoke from coming out; by this Means the Milk would all be soon heated, with any kind of Fuel, and that with much less in Quantity than in the common Way.

60. And the more effectually to prevent the Smoke from coming at the Milk, it may be well to have the broad outer Rim of the Pans turned perpendicularly downwards, three or four Inches, that it may enter deep into a circular Groove of Sand; and if it shall be needful the Sand may be wetted in order the more effectually to prevent the Passage of the Smoke: I thought of this Method about fifty Years since on tasting the smoky Butter in *Somersetshire*. By the same Means the Poor might save much Fuel in boiling the Pot, especially in Summer, when a Fire is wanted only for boiling the Pot.

61. When any Pans are to be removed from the Stoves, the Ascent of the Smoke thro' the uncovered Stove, may be prevented by first closing the Flue near the Fire, by an Iron Sliding-shutter or Register.

62. Milk might thus most commodiously be heated to a scalding Heat with little Fuel, fit for Ventilation, in a Vessel of a proper Depth, set in the same Manner as the Pans in a Stove, to secure it from Smoke, with Bellows fixed properly near it: (see *Fig. 3*.) By this Means there would be little Trouble or Expence in curing ill tasted Milk by Ventilation.

63. *May* 14th, meerly to see what the Event would be, a Gallon of new Milk, just from the Cow was ventilated, for an Hour and half, which produced six Ounces of Butter; and tho' it was ventilated half an Hour longer, yet no more Butter was made; it was whitish, wanting both the Colour and Taste of good fresh Butter.

64. I am credibly informed, that in the Places famous for making the best fresh Winter Butter, they set the Pot of Cream in warm Water, so long as till it has acquired that small Degree of Sourness, which it very soon has in warm Summer Weather, which gives it its agreeable Flavour. And in order to give it Colour, they grate a well coloured Carrot into a little Milk, which as soon as stained, is strained from the Carrot thro' a Sive, and then mixed with the Cream.

which it would otherwise require to refill with cold Liquor, and bring to a distilling Heat. This great Advantage, added to that of distilling about double the Quantity by Ventilation, must needs both together be of so great Improvement in distilling three Parts in four more than usual, as will be of the greatest Benefit to Navigation in several respects. By this means the Still need be emptied and cleansed from the Chalk, and very salt Water, only at the end of each daily Distillation.

80. And whereas if all the Chalk requisite for a whole Day's Distillation were put into the Still at once, it might be suspected that such a Quantity laying at the Bottom of the Still, might hinder the boiling of the Water: I put into a broadly flat-bottomed sauce-pan, half an Inch Depth of powdered Chalk, and two Quarts of Water boiled in ten Minutes: The Water mingling freely, especially in the time of boiling, among the thin pappy Chalk, was freely acted on by the Fire. But as in boiling it frothed much, it may be proper to put at first into the Still only as many half Ounces of Chalk as there are Gallons of Water; and afterwards from time to time in proportion to what additional Water shall have run into the Still, at a Hole purposely made in the Head of the Still, taking care to ventilate while the Chalk is putting in, which will cause it to mix intimately with the Water, and thereby have the better Effect on it. And perhaps a less Proportion of Chalk than half an Ounce to a Gallon of Sea-water may suffice.

81. The Degree of Fulness or Emptiness of the Still, may be known by putting a small floating metalline Vessel, in at the Chalk Hole, with a long slender Wire to it.

82. Mr. *Wyche* thought of other Means the better to promote Distillation, *viz.* by conveying the Air from the Bellows through a Pipe with several spiral Coiles, thereby in passing thro' boiling Water in the Still, to make the ascending Showers of the Air the hotter, and so the less refrigerating. Also to have the warm Water pass from the Worm-tub, in a Pipe, with several the like spiral Coiles, in the Head of the Still; thereby the more to heat the entering Water. But we find, by the above-mentioned Experiments, that there is no occasion for either of them; but if wanted may be used.

83. As to the Benefit of Ventilators in preserving the *Health and Lives* of People in Slave and other Transport Ships, I have received farther Confirmation of their great Benefit, in a Letter from Dr. *Demainbray,* who has shewn Courses of Experimental Philosophy to his Royal Highness the *Prince* of *Wales* and *Prince Edward,* viz. "That in the Year 1753 Ventilators were put into the Vessels in the Slave Trade at *Bordeaux,* and in other Ports of *France;* the happy Effect of which was, that instead of the Loss

of one-fourth of those valuable Cargoes, in long Passages from *Africa* to the *French* Plantations, the Loss seldom exceeded a twentieth. And since my Return to *England*, I have been informed of a *French* Vessel, which by this self-evidently reasonable Precaution, saved 308 out of 312 Slaves, spite of most tedious Calms and a long Passage." And Dr. *Garden*, in his Letter to me, dated *Charles-Town, South-Carolina, March* 24th, 1756, says, *viz.* "It is indeed very wonderful, that the Slave Merchants do not come into the Use of Ventilators. There are few Ships come here from *Africa*, (even though they call for fresh Provisions and Water at the *Islands*) but have had many of their Cargoe thrown overboard; some one-fourth, some one-third, some lose half; and I have seen some that have lost two-thirds of their Slaves. I have often gone to visit these Vessels on their first Arrival, in order to make a Report of their State of Health to the Governor and Council; but I never yet was on board one, that did not smell most offensive and noisome. What from Filth, putrid Air, putrid Dysenteries (which is their common Disorder) it is a wonder any escape with Life."

84. A probable Means lately occurred to my Thoughts, tho' not perfectly to cure, yet much to abate the great Degrees of stinking of the Bilge-water in the Well of Ships, *viz.* by laying at the Bottom of the Bilge-water, round the Main-mast, Copper Pipes full of very small Holes, not one twentieth of an Inch in Diameter; and blowing Showers of Air up thro' them from small Black-smith's Bellows, fixed out of the way, within the Well. Such Bellows may be worked with great Ease, for one, two, or three quarters of an Hour, in every 24 Hours, more or less, as shall be found requisite by Experience: But with this Precaution, that whenever the Water stinks much, first to pump it out of the Ship, and after letting in sweet Water, then to ventilate it a little now and then, as shall be found needful; but care must be taken not to ventilate very stinking Water, because it may increase the Unwholsomeness of the foul Air in Ships. By this means Ships will probably be made something less unhealthy; for the putrid Vapours which arise from stinking Water, must needs greatly contribute to increase the Putridness, and consequently the Noxiousness of the foul Air in Ships. This Precaution therefore, with Ventilators frequently to convey off the close foul Air, will be effectual Means to make Ships much more healthy; as will the changing the foul Air in Goals, Hospitals, and sick Rooms, with proper Precautions, make them more wholsome: And it is with pleasure that I observe, that these salutary Means are coming more in use in Hospitals, *&c.* either by means of Ventilators, or by admitting fresh Air in small Quantities, and those spread in thin Sheets, so as not to have the entering Air blow directly on, and incommode the Patients. Where it can be had, a thorough Air entering on one side, and passing out on the other side of Wards or Rooms, will be

best; because the Change will be almost constant, and may be as gradual and gentle as we please. I shall give a more particular Account of this in my second Volume of Ventilators.

85. As to the curing the ill Taste of Milk from the Food of Cows; and also curing musty Liquors, I made the following Trials, *viz.* When in the Beginning of *March* the Turnips had made large Shoots, and were thereby become very rancid, two Cows being fed with Turnips only for seven Days, their Milk had a very disagreeable Smell and Taste. On ventilating it scalding hot, with ascending Showers of Air, at first the ill Smell increased, but in two Minutes that Smell was much abated; and with five Minutes Ventilation, there was only the common Smell of good Milk; which shows that the rancid Oil of Turnips, which gives the ill Taste, is very volatile: After ten Minutes Ventilation, there was no ill Taste or Smell; and it was the same after 15 and 30 Minutes Ventilation. By this means therefore the ill Taste of Milk from some Food of Cows, may easily be cured. Experience will show what Degrees of Ventilation will be requisite for larger Quantities of Milk; as also for curing the ill Tastes from different Kinds of Food; as also from the shorter and longer Times of feeding on such Foods. It is observable that the Breath of these Cows was disagreeable; whence we see how freely contagious Infections may be conveyed through the Lungs, from tainted putrescent Blood.

86. A Cow having been fed for sixty Hours with a good Quantity of Crow Garlick, mixed with cut Grass, towards the end of *June*, the Milk had a very disagreeable Smell and Taste; which was not cured, though something better, after 30 Minutes Ventilation, while scalding hot. I repeated the same Ventilations for 15 Minutes, with the two following Evening and Morning Milks of the same Cow, *viz.* 12 and 24 Hours after the Cow had left off eating *Crow-Garlick*; at which times the ill Taste and Smell was sensibly abated before Ventilation, and something more after it, yet was far from being cured. Hence we see that the ill Taste and Smell of the *Crow-Garlick* Milk, cannot thus be cured, tho' somewhat amended. The ill Taste of this Cow's Milk continued for about five Days after she had left off eating of the *Crow-Garlick.* And perhaps where Cows have eaten but a small Quantity of it, the Abatement of the ill Taste may be so considerable, as to recompence the Trouble of ventilating it. As Opportunities offer, I will make the like Trials, with other ill-tasted Milks, such as that from autumnal Leaves, *&c.* and hope that others will do the like.

87. Cream or Milk Sillabubs may most commodiously and easily be made in Plenty, in a few Minutes, by means of a small Tin Air-box, three Inches in diameter, and three quarters of an Inch deep; the flat Bottom of

the Pot, which contains the Cream or Milk, to be but little wider than the Air-box, that the Air from the Box may the better come at it: But the wider and deeper the upper Part of the Pot is, so much the better, for the Froth expands much.

88. Scalding-hot musty Vinegar was cured by repeated Trials, of near three Gallons in each Trial, by ten Minutes Ventilation. And giving Vinegar a scalding Heat, does not weaken or damage it, as it does Wine, whose vinous Spirit is carried off both by Heat and Ventilation; for when some strong musty Raisin Wine was ventilated hot, it was cured in five Minutes: But the vinous Spirit, which strongly affected the Nose in flying off during Ventilation, being gone off, the vapid Wine would not kindle into a Blaze, when thrown into the Fire, as it did to a great Degree before it was heated and ventilated. Neither musty Wine nor Vinegar were cured, tho' something bettered by 30 Minutes Ventilation, when cold.

89. Mr. *Jones*, a Chemist in *Cranbourn-Ally, Leicester-Square*, ventilated a Gallon of Proof Malt Spirits 15 Minutes cold, in which time it wasted two Ounces and half; whereas a like Quantity of common cold Water wasted but half an Ounce in 15 Minutes Ventilation, *viz.* but one-fifth part of what the Spirits did; and the same Spirits ventilated hot, wasted no less than 5 Ounces in 5 Minutes; which Ventilation made them sensibly better tasted than the unventilated. But the great Waste shows that these volatile vinous Spirits, ought not to be ventilated, neither hot nor cold. Besides that, the 15 Minutes cold Ventilation had but little Effect in bettering the Spirit.

90. In order to know whether Fish in a Vessel of Water would live the longer for having Showers of Air blown up thro' the Water, *May 25, 1756*, at seven in the Morning, the Wind N. E. the Mercury in *Farenheit*'s Thermometer, 50 Degrees, and it continued so cold, that at one o'clock it rose but to 60 Degrees; I put twelve Dace into a Pail *A* in two Gallons of fresh Pond-Water, thro' which a Stream ran, and twelve more into a Pail *B*, with the like Quantity of Water; one of which Fishes in *B* was sick, as appeared by turning its Belly upwards; as were also two Fishes in the Pail *A*, which was occasioned by being all brought in a Pail above half a Mile from the *Thames*.

91. At 45 Minutes past 8, most of the Fish in the Pail *B*, were turned Belly upwards, and lay as dead; at 30 Minutes past 9 seven of them were dead; at 30 Minutes past 11 all but three were dead; at two o'clock but two Fish remain alive in *B*; which remained alive, tho' sick, at ten that Night, *viz.* at the end of 14 Hours.

92. The good Effect of blowing every quarter of an Hour, with 25 Strokes each time with double Bellows, Showers of fresh Air up thro' the

Water, was, *viz.* they all continued well, and the larger of the two sick Fish recovered; but the lesser died at 4 o'clock, *viz.* at the end of nine Hours; upon each Ventilation it turned its Back upwards; but soon after the Ventilations, it constantly fell precipitate with its Head foremost, to the Bottom, and there turned Belly upwards. After 4 o'clock the ventilated Water frothed with larger Bubbles, this owing to the Slime of the Fish. The last Ventilation was at 10 at Night, when the Fish in the ventilated Water were well, and would probably have long continued so by the Salutariness of Ventilation; but that being discontinued, they were all found dead the next Morning, except one which had some small Degree of Life.

93. *June 7*, the Wind S. W. cloudy, the Thermometer at 58 Degrees, 13 live Gudgeons were put into two Gallons of fresh Pond-water in a Pail *A*; and a like Number into a Pail *B*, at 10 Minutes before 7 in the Morning. At 50 Minutes past 7, two in *B* began to be sick; at 8, half of them came up for Air, and showed Uneasiness; at 15 Minutes past 8 two are dead; at 30 Minutes past 8, eight more turned Belly upwards; at 8 Minutes past 9, five are dead in *B*, and five more sick; at 30 Minutes past 9, seven are dead, and four sick, two well; and 30 Minutes past 10, eight are dead; at 11 two only alive, and also well, though they showed some Uneasiness by their raising their Mouth to the Surface, which they continued to do till nine, when they were taken out of the Water. Hence we see, by this, and the preceding Experiment on Dace, that one or two Fish may be kept alive many Hours longer than a greater Number can be, by means of the small Portion of fresh Air, that is continually mixing with the Water, on which it presses.

94. The Water in the Pail *A* was ventilated from 10 Minutes before 7, to 6 in the Evening, by blowing every quarter of an Hour Showers of fresh Air up thro' it, with 25 Strokes of the Bellows; by which means the Fish continued all well, laying quiet at the Bottom: At six we ceased to ventilate; for an Hour and half after which, there was no Signs of Sickness, after two Hours two of them showed Signs of Uneasiness; and at nine most of the Fishes turned Belly upwards, and lay at the Bottom dead or dying. By comparing this Event with that at eight in the Morning, we see there is more Air in this ventilated Water than in the Pond-water, as is probable by the Gudgeons living longer in it without Ventilation, than in the Pond-water. And accordingly the specific Gravity of unventilated Pump Well-water, was a very small Matter greater, than that of the same, after being ventilated with 100 Strokes of the Bellows, as I found by the Hydrometer; and there was nearly the same Difference between the specific Gravity of unventilated and ventilated Table Beer.

95. Tho' from these Experiments it is manifest, that Fish die for want of constant Supplies of fresh Air in the Water, yet when taken out of the

Water very lively, they soon die, notwithstanding the Surfaces of their Gills are then exposed to the immediate Contact of the Air; which shows that either the Air does not enter from the Gills to the Blood, from the open Air, as it does from the Water; or that the Circulation of the Blood is stopped by exchanging their proper Element, Water, for Air; as the Circulation of the Blood of Land Animals is soon stopped by immersing from Air into Water.

96. Hence we see the Benefit of frequently replenishing the Water with fresh Air, which we find is necessary not only to preserve the Life of Land Animals, but also of Fish; as also the Use of their Gills, to spread in thin Sheets fresh Supplies of Water, that they may the better come at the Air in the Water; for which Purpose both Sides of their Gills are furrowed with many fine Furrows, not only thereby to enlarge their Surfaces, but also more minutely to divide the Water, whereby to come at the Air in it.

97. Whence we may reasonably infer, how requisite it is, in order to keep the Blood in a salutary State, to have almost constant Supplies of the Breath of Life, fresh Air, to mix with it: For if the principal Use of the Gills were only to cool, and churn, and comminute the Blood, Water devoid of Air could as well perform that Office, as Water repleat with fresh Air. It must therefore be of Importance for all Animals to have so necessary a vital Fluid fresh and pure, and not foul and putrid.

98. A Fisherman informing me that the Fish were apt to die in the Well-boats, it occurred to me that it might, in a good measure, be remedied, by fixing upright a Board of a proper Breadth at the Outside of the Boat, as long as the Well is deep, on that side of the Well which is next to the Stern, and opening at an Angle of about 45 Degrees towards the Head; which, by checking the Course of the Stream of the River near the Boat, will raise the Water a little, and thereby cause it to run into the Well on that Side, and out on the other Side of the Boat; and the Run of the Water thro' the Well, will be more accelerated, if a like Board is fixed in the same manner on the other side of the Boat at the Side of the Well next to the Head of the Boat, but opening towards the Stern at an Angle of 45 Degrees, by which means the Water will be lowered on this, as much as it is raised on the other Side of the Boat.— —Another Means to have a brisker Current of Water thro' the Well, would be to moor the Boat in a fixed Position across the Stream when it can conveniently be done. I am informed that there are Passages from Head to Stern, for the Water to pass freely thro' the Fish-Well of some Vessels. We see, from the Event of the above Experiments, the Importance of frequently changing the Water in the Fish-Well of Ships, especially when there are many Fishes in it. Hence also we see how salutary it is to Fish, to have the Surface of the Water agitated into Waves by Winds, whereby Plenty of fresh Air enters, and is blended with the Water.